MY INVENTIONS

New Renewable Electricity Sources

1. *Atmospheric pressure machine*
2. *Hydrostatic machine*
3. *Frigorific cycle atmospheric machine*
4. *New protective device against current leaks without grounding*

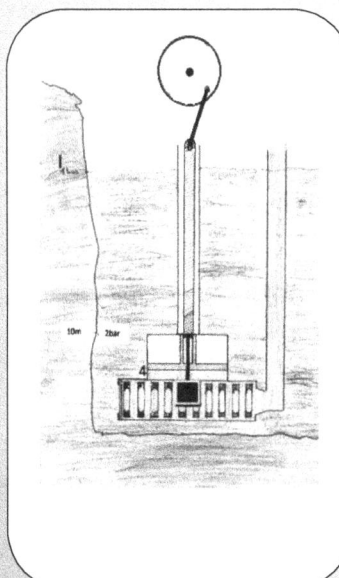

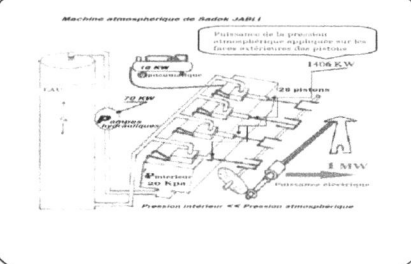

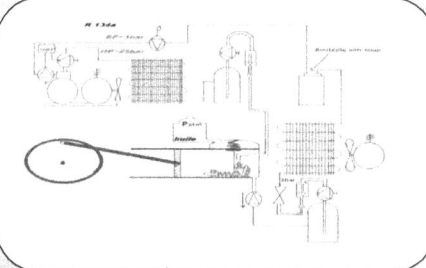

MY INVENTIONS
New Renewable Electricity Sources

This book contains the descriptive documents of my inventions with all scientific and technical information proving the possibility of using atmospheric pressure and hydrostatic pressure as a new sources of renewable energy.

I gave up the industrial property rights of these inventions to contribute to the development of the field of renewable energy and to open new horizons to research in this field, which I consider as the only haven for humanity to get rid of water and food shortage problems.

__1.__ Invention of a new technology to produce energy by atmospheric pressure: __Atmospheric Machine__

__2.__ Invention of a new technology to produce renewable energy by the means of water hydrostatic pressure: __Hydrostatic machine__

__3.__ Invention of an atmospheric machine using frigorific cycle to produce electric energy: __Frigorific cycle atmospheric machine__

__4.__ The book also contains my invention "__New protective device against current leaks without grounding__" intended to minimize the risks resulting from the disadvantages of earth leakage circuit breakers and neutral systems to protect millions of users around the world who do not have an earth ground connection where is the greatest number of victims of electric shocks.

I would like to invite all institutions and researchers interested in this field to continue researching these systems.

We can produce renewable energy from everything around us, such as atmospheric pressure, hydrostatic pressure and gravity.

You can use these researches to your own profits (final course projects, theses, development, construction...) without any legal problem. I just want these new sources to be used to generate renewable energy.

1. Atmospheric pressure machine

This is a description of my new invention « Atmospheric pressure machine » with all scientific and technical information proving the possibility of using atmospheric pressure as a new source of renewable energy.

The descriptive document includes:

-The 1st part: Invention description with regard to theoretical aspect, and scientifically proving the possibility to produce a continuous mechanical movement by the means of atmospheric pressure.

-The 2nd part: description of an invention related to the practical aspect, it is about the development of the first invention. This description shows the practical aspects of the invention use and includes all required technical and scientific data.

The atmospheric pressure machine is an atmospheric turbine that converts kinetic energy of the cylinders whose external sides are under the atmospheric pressure into mechanical work.

Work and energy principle:

The input energy is the atmospheric pressure on the cylinders (converted into kinetic energy, the output energy is the electric energy, and the efficiency rate is 0,7.

2. Hydrostatic machine

This invention is related to an hydrostatic machine submerged in water at a determined depth (10m, 20m, or deeper) which is used to produce renewable electric energy from the energy transferred by the

action of hydrostatic pressure on the piston. It is a new technology to produce renewable energy at a very high conversion rate, totally independent from location and climate, at a very low manufacturing cost compared to other renewable energy sources.
The document includes all required scientific and technical information.

3. Frigorific cycle atmospheric machine
This invention consists of an atmospheric machine using frigorific cycle to produce renewable electric energy out of atmospheric pressure energy coming from nature and applied on the piston(s) external face. For this machine, ambient air pressure is the origin of the system driving force.

4. Invention of a new protective device against current leaks without grounding
This invention is related to a new protective device against indirect contact (accidentally energized metallic body) without grounding intended to minimize the risks resulting from the disadvantages of earth leakage circuit breakers and neutral systems to protect millions of users around the world who do not have an earth ground connection where is the greatest number of victims of electric shocks

ATMOSPHERIC PRESSURE MACHINE

Introduction:

This is a description of my new invention « Atmospheric pressure machine » with all scientific and technical information proving the possibility of using atmospheric pressure as a new source of renewable energy.

The descriptive document includes:

-The 1st part: Invention description with regard to theoretical aspect, and scientifically proving the possibility to produce a continuous mechanical movement by the means of atmospheric pressure

*-The 2nd part: description of an invention related to the practical aspect.
It is about the development of the first invention. This description shows the practical aspects of the invention use and includes all required technical and scientific data. It is through this invention that suitable equipment is targeted to manufacture atmospheric pressure machines.
I invite you to discover this Energy Breakthrough. To understand this Machine functioning principle, remember the following:*
<u>*Nota*</u>*:*

The atmospheric pressure machine is an atmospheric turbine that converts kinetic energy of the cylinders whose external sides are under the atmospheric pressure into mechanical work.

Work and energy principle:

The input energy is the atmospheric pressure on the cylinders (converted into kinetic energy, the output energy is the electric energy, and the efficiency rate is 0,7.

Invention

Atmospheric pressure machine
(Theoretical aspect)

I. **Preliminary**

This invention is related to a machine that uses atmospheric pressure as a renewable energy source.

The proposed machine uses atmospheric pressure to produce a mechanical movement.

The proposed invention is based on two simple observations:

1st Observation :

Let us take the following syringe example

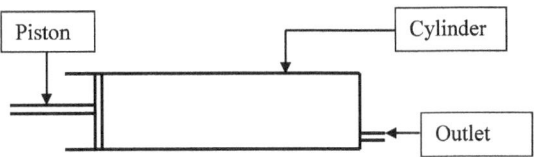

- When the outlet is locked, if the piston is pulled, then released, the piston quickly comes back to its initial position owing to the internal pressure decrease.

⇒ The piston movement is necessarily due to a force $\vec{F}$ coming from the difference between the atmospheric pressure and the internal pressure, and as

$P_{atm} > P_{int}$

$$\Rightarrow (P_{atm} - P_{int}) \times S_p = F$$

2nd Observation

- The syringe outlet is blocked.
- The piston is pulled ⇒ pressure decreasing inside the syringe.
- While holding the piston pulled, the syringe is immersed into a liquid.
- The outlet is opened.

⇒ The liquid flows into the cylinder until the empty volume is filled.

Starting from these two observations, I imagined building a machine made of two cubes; each cube is linked to a cylinder. The two cylinders contain the same piston, or liked to a crank shaft.

- Each cube has two openings, one is linked to a basin containing a liquid, and the other is linked to a closed vacuum tank, its internal pressure being very small compared to atmospheric pressure.
  The set "Cube – Cylinder" is called intermediate chamber.
- A monitoring system is used to activate the two intermediate chambers openings by activating the piston under the pressure difference between its two ends : $\Delta P = P_{atm} - P_{int}$
- The four openings monitoring is synchronized.
- A hydraulic pump is connected to the tank to clear out the liquid quantity suctioned at every cycle movements.
- The monitoring system is synchronized as follows :

1cycle/ second $\Leftrightarrow Q_{Vpump} = 2 \times V_{\text{intermediate chamber}} / S$

- Piston cannot be left working spontaneously alone because no work can be won.

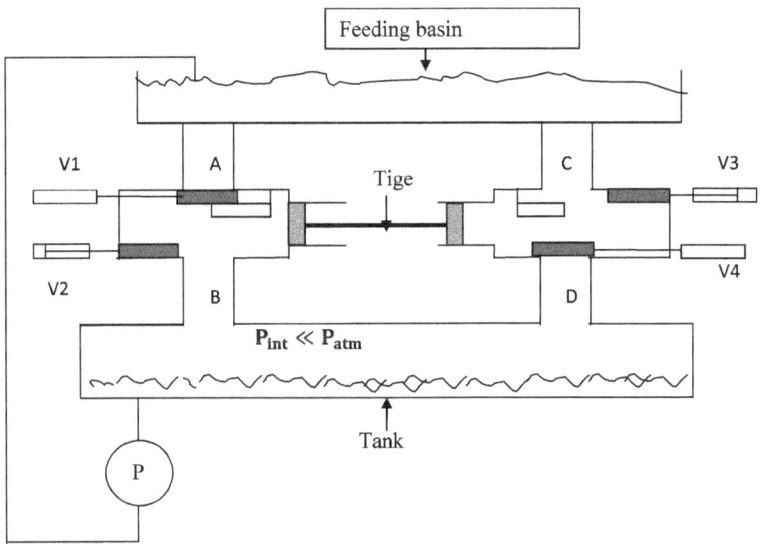

Machine diagram (theoretical point of view)

Note: both pistons are tied to a rod or a crank shaft.

II. Operation description

The machine consists of :

- Feeding basin containing a fluid, liquid for instance.

- A closed vacuum tank, its internal pressure is very small compared to atmospheric pressure but higher than saturation pressure of the used liquid.
- A hydraulic pump connected to the tank has to clear out the liquid to the feeding basin.
- Two intermediate chambers.

1. Operation

The basin and the tank are linked by two intermediate chambers.

The intermediate chamber consists of a pipe with two openings linked to a cylinder.

- The upper opening is connected the feeding basin.
- The lower opening is linked to the tank.
- The two pistons of the two cylinders are linked by the same rod or by a crank shaft.
- The two intermediate chambers are operated through a monitoring system that activates the four openings.
- The monitoring system consists of 4 compressed air cylinders and a compressor.
- The machine operates whenever the four openings are activated as follows :

  ❖ **Initial stage :**

  A: closed   AND   C: Opened

  B: opened   D: closed

  ❖ **Intermediate stage :**

  A: closed   AND   C : closed

  B: closed   D : closed

  ❖ **Final stage :**

  A: opened   AND   C : closed

  B: closed   D : opened

Thus, every time when we change the state of the openings according to the three stages mentioned here above, one of the two pistons is found under the pressure difference: $\Delta P = P_{at} - P_{int}$, while the other piston is left free.

The intermediate stage role is to avoid the liquid to be directly suctioned between the tank and the feeding basin.

At any change in the openings state, a movement is generated as a rod translation or a crank shaft rotation.

- The two pistons movement has to be synchronized and should never be spontaneous.

That means:

The monitoring system changes the openings states twice a second (2 times / second), which makes a cycle per second (1 cycle / Second).

- The hydraulic pump returns the water volume given at every cycle from the tank to the feeding basin. The flow is $Q_v = 2\ V_{ch}/s$

Note :

Q_v : **pump volumetric flow**

V_{ch} : **Intermediate chamber volume**

If another solution is chosen where the piston moves, the pump work becomes higher than pistons one, and no work is won by such a machine.

Thus: the only way to win work is to synchronize the openings monitoring system.

2. Monitoring system operation:

The monitoring system consists of 4 compressed air cylinders and a compressor.

The compressed air cylinders are timed to have one cycle per second.

Monitoring system diagram from a theoretical point of view

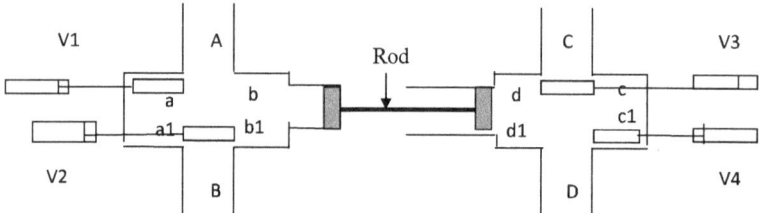

V1, V2, V3, V4 : compressed air cylinders

a, b, a1, b1, d, c, d1, c1 : position sensors

The following GRAFSET explains the monitoring system operation:

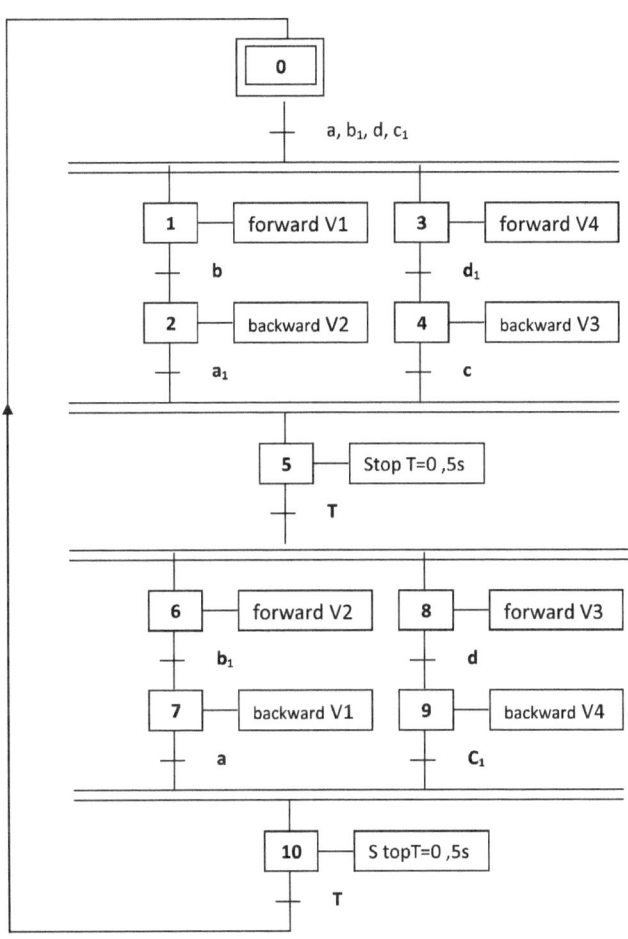

III. Required conditions for good operation

- Volume of the two intermediate chambers should be small compared to the feeding basin volume and to the tank volume.

$V_{ch} \ll V_{tank} \Rightarrow P_{int} \cong$ **constant**

$V_{ch} \ll V_{basin} \Rightarrow$ **flow continuity**

- Internal pressure should be very small compared to atmospheric pressure.
- Internal pressure should be higher than the used liquid saturation pressure.
- The monitoring system should be synchronized to get one cycle per second.
- Before vacuum settling in the tank (pressure decrease), a liquid level limit should be determined and maintained by the pump.
- The openings and the pistons must have the same diameter.

IV. **Machine work calculation :**

1. Work of the hydraulic pump

The pump is located under the tank. It must elevate liquid from the tank to the feeding basin which is at 5 m level.

Pressure loss is estimated at 0,1 m

The used liquid is water, volumic mass $\rho = 1000 \; Kg/m^3$

Let us apply Bernoulli formula to the following system:

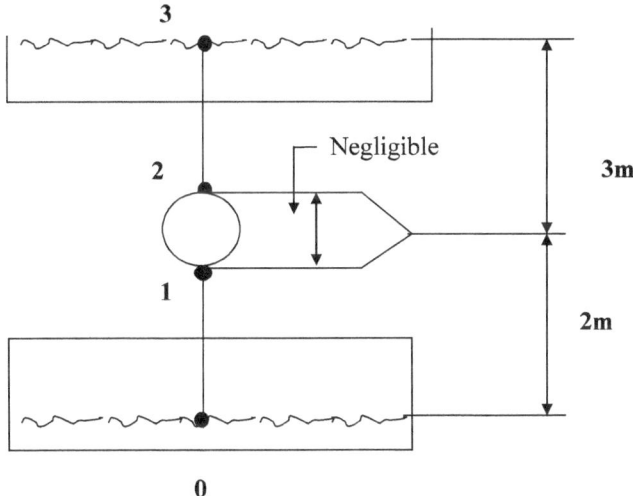

$$\omega_{0-3} = \frac{P_3 - P_0}{\rho} + \frac{1}{2}(C_3^2 - C_0^2) + g((Z_3 - Z_0) + J_{0-3})$$

$$P_3 = P_{atm} = 101{,}325 KPa = 10^5 \; Pa$$

$$P_0 = 20 KPa = 2\;10^4 \; Pa$$

$$C_3 = C_0 = (\text{motionless fluid outside the pipe})$$

$$J_{0-3} = 0{,}1 \times 5 = 0{,}5 \text{m of fluid}$$

That is added to $Z_3 - Z_0$

Then $\omega_{0-3} = P_3 - P_0 + \frac{1}{2}\rho(C_3^2 - C_0^2) + g\rho((Z_3 - Z_0) + J_{0-3}) = 133900$ J

Then $\omega_{\text{hydraulic pump}} = 133900$ J $= 133.9$Kj

2. Work of the monitoring system

As a consequence of the technological solution used, the work accomplished by the monitoring system is very weak compared to the piston work. That is because the force applied to the surface locking the opening is perpendicular to the piston one.

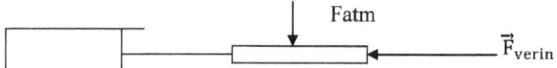

Conclusion :

$$\omega_{\text{piston}} \gg \omega_{\text{pump}}$$

The work gain is obtained owing to the flow discontinuity caused by the monitoring system synchronization which makes that pump work is low compared to the two pistons work.

1- Dimensioning example and machine power calculation

1. System study (Tank+piston cylinder).

Let us suppose the following system

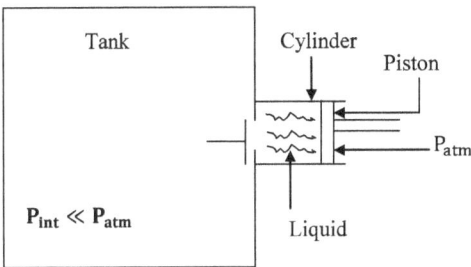

The system is under the following conditions:
- P_{int} in the tank is very small compared to atmospheric pressure.
- The fluid saturation pressure as function of the temperature is lower than the internal pressure inside the tank in such a way that the fluid remains as a liquid.

$S_A > S_B$ et $f \cong 0$

S_A : Piston section

f : Coefficient of friction

S_B : Section of the opening

Bernoulli formula :

$$\frac{\rho V_A^2}{2} + \rho g Z_1 + P_A = \frac{\rho V_B^2}{2} + \rho g Z_B + P_B$$

$$\Leftrightarrow \frac{1}{2g}(V_B^2 - V_A^2) = \frac{(P_A - P_B)}{\rho g}$$

$$\Leftrightarrow V_B^2 - V_A^2 = \frac{2(P_A - P_B)}{\rho}$$

As we have $V_B S_B = V_A S_A \Rightarrow V_A = V_B \frac{S_B}{S_A}$

$$V_B^2 - (V_B \frac{S_B}{S_A})^2 = \frac{2(P_A - P_B)}{\rho}$$

$$V_B^2 \left(1 - (\frac{S_B}{S_A})^2\right) = \frac{2(P_A - P_B)}{\rho}$$

With $S_A > S_B \Rightarrow (\frac{S_A}{S_B})^2 < 1$

$$V_B^2 = \frac{2(P_A - P_B)}{\rho \left(1 - (\frac{S_B}{S_A})^2\right)}$$

$$V_B = \sqrt{\frac{2(P_A - P_B)}{\rho \left(1 - (\frac{S_B}{S_A})^2\right)}}$$

And $V_A = \frac{V_B S_B}{S_A}$

2. The two pistons power

Considering the piston diameter d= 20cm = 0,2m.
The liquid is water, volumic mass $\rho = 1000 Kg/m^3$.

$$S_{piston} = \frac{0,2^2}{4} \pi = 3,14 \, 10^{-2} m^2$$

$$P_{int} = 20Kpa = 20 \, 10^3 Pa$$

$$P_{atm} = 101Kpa \cong 10^5 Pa$$

$$F = (P_{atm} - P_{int}) \times S_{piston}$$

Numerical application : $F = (10^5 - 20 \times 10^3) \times 3,14 \, 10^{-2} = 2512 \, N$

It is supposed that the opening section is slightly lower that the piston section. According to Bernoulli theorem, the piston speed is calculated taking into account an opening diameter of 19 cm:

$$V_B = \sqrt{\frac{2(P_A - P_B)}{\rho\left(1 - (\frac{S_B}{S_A})^2\right)}}$$

With $S_{piston} = \frac{0.2^2}{4}\pi = 3,14 \ 10^{-2} m^2$

$S_B = \frac{0,19^2}{4}\pi = 2,83 \ 10^{-2} m^2$

Numerical Application $V_B = \sqrt{\frac{2 \times 80 \times 10^3}{0,85 \times 10^3 \times 0,2}} = 30.67 \ m/s$

$$\Rightarrow V_p = \frac{S_B}{S_A} V_B$$

Numerical Application : $V_p = 30.67 \times 0,8 = 24.5 \ m/s$

Then, the piston mechanical power is:

$$P_m = F \times V_p$$

NA: $P_m = 2512 \times 24.5 = 61544 W = 61.544 \ KW = 83.9 \ Cv$

Then, the mechanical power for a movement cycle is:

$$P_{m \ for \ 2 \ pistons} = 2 \times P_m$$

AN: $P_{m \ for \ 2 \ pistons} = 167,8 \ Cv = 123 \ KW$

3. Power of the hydraulic pump

The hydraulic pump has to elevate water that has a volumic mass of:

$$\rho = 1000 \ Kg/m^3$$

The pump flow is: $Q_V = 2 \times V_{intermediate\ chamber}/S$

$V_{intermediate\ chamber} = V_{cube} + V_{cylinder}$

$V_{cube} = (0,4)^3 = 0,064\ m^3$ et $V_{cylinder} = 15,7 \times 10^{-3} m^3$ ⇒
$V_{intermediate\ chamber} = 79,7\ 10^{-3} m^3 = 79,7\ \ell$

It is supposed that the pump flow is $Q_V = 80\ \ell/S$ and we have

$W_{pompe} = 133900\ J$

Then $P_{m\ pump} = w_{pump} \times Q_{massic} = w_{pump} \times Q_V$

$P_{m\ pump} = 133900 \times 80 \times 10^{-3} = 10712\ W = 10,712\ KW = 14,5\ Cv$

4. Power of the monitoring system

A 5.5 cv compressor can be used.

5. Power of the machine

The gained power by the machine is:

$P_{machine} = P_{m\ for\ 2\ pistons} - (P_{pump} + P_{system})$

Avec $(P_{pump} + P_{system}) = 20\ Cv$

AN: $P_{mach} = 167.8 - 20 = 147.8 Cv = 108.7\ Kw$

Development of the Atmospheric Pressure Machine
(Practical aspect)

I- Presentation :

This invention is related to the development of the previously invention entitled « Atmospheric pressure machine ».

In the previously machine, it is the atmospheric pressure that provides the work to the two pistons which are linked to a rod or a crank shaft, and they are displaced whenever the vacuum attracts one piston while the other one is left free

The development concerns :

- A three stroke monocylindric atmospheric pressure machine,
- A three stroke multi-cylinders atmospheric pressure machine ,
- An atmospheric micropower plant,
- An atmospheric power plant.

II- Operation description :

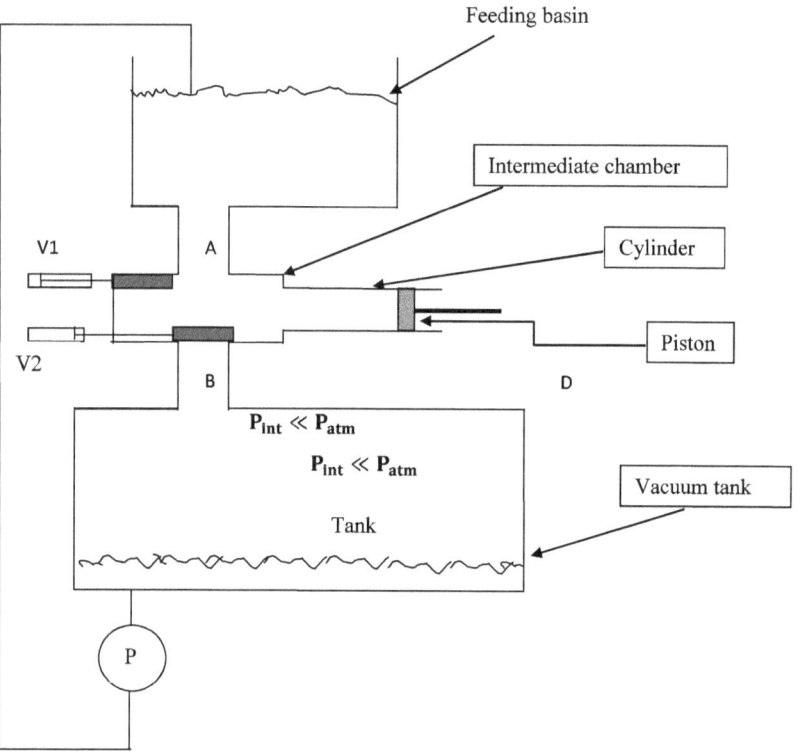

The machine consists of :

- Feeding basin containing liquid,
- A vacuum tank where the internal pressure is very small compared to atmospheric pressure, but it is higher than the used liquid saturation pressure,
- One or more intermediate chambers according to the machine type (monocylindric or multicylindric),
- The intermediate chamber is made of a parallelepiped rectangle having two openings connected to a cylinder equipped with a piston.
- The upper opening is connected to the feeding basin and the lower opening is connected to the vacuum tank.

Intermediate chamber

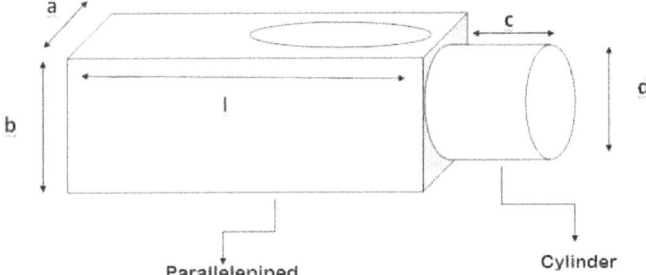

Parallelepiped　　　　　　　　　　Cylinder

- A monitoring system is used to operate the intermediate chamber in a manner that the piston is connected to the feeding basin and to the vacuum tank according to a three stroke operation cycle (one cycle per second).
- The monitoring system consists of compressed air cylinders, position sensors, and a compressor to open and lock the openings according to a synchronized sequence in such a way that the vacuum attracts the piston (or the pistons set) once a second under the effect of the pressure difference ($\Delta P = P_{atmospheric} - P_{internal}$) which acts on the external face of the psiton (or the the pistons set).
- A hydraulic pump connected to the tank has to clear out the liquid dropped at every cycle to the feeding basin.
- The pump selection is made to fulfill the flow requirement where the pump flow equals the water volume going to the tank at every cycle : $Q_V = n * V_{Chambre\ intermédiaire}/s$. where Q_V : volumic flow, n : number of intermediate chambers in the machine
- The operating cycle of the atmospheric pressure machine is a three stroke one:

* First stroke : intake

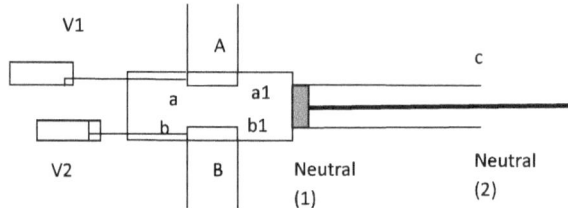

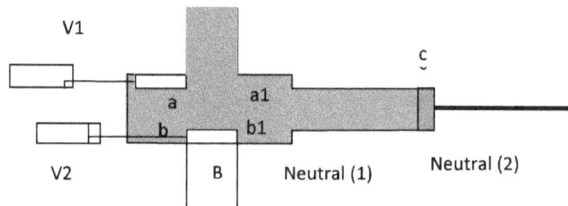

> - Intake slide opening
> - Piston displacement from neutral point (1) to neutral point (2)
> - Liquid filling the cylinder

* Second stroke : synchronizing time.
  > - Stopping time T= 1s (one second)
  > - Piston free displacement back and forth between neutral point (1) and neutral point (2) due to the flywheel or to the crank shaft linked to the piston (free piton).

* Third stroke : suction time (driving time)

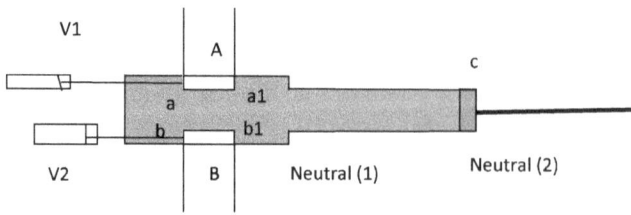

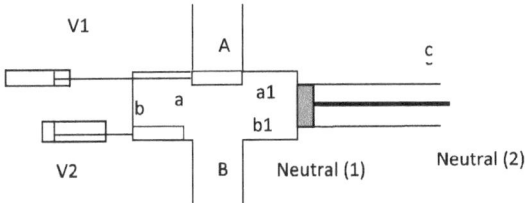

- Piston location at neutral (2) (position sensor).
- Locking the intake slide.
- Opening the suction slide.
- Piston displacement from neutral (2) to neutral (1).
- Locking the suction slide.
- Opening the intake slide.

- The following GRAFSET explains the operation of one intermediate chamber in the monitoring system.

  Given:
  a, a_1 : location sensors of the cylinder V_1
  b, b_1 : location sensors of the cylinder V_2
  c : Presence sensor of the piston at neutral point (2).

GRAFSET

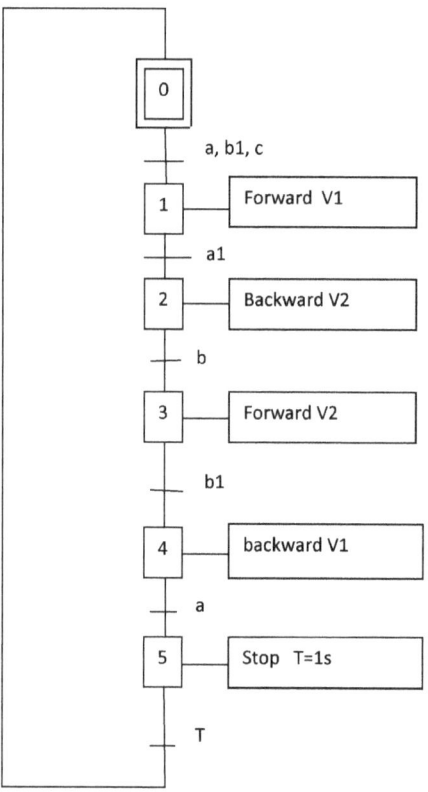

⟹ **Note**: The stop time (synchronization done through the sensor « c » which is activated one a second.

- Required conditions for good operation :
  - The volume of the intermediate room must be small compared to the feeding basin volume and to the tank volume.
    $n*V_{chamber} \ll V_{tank} \implies P_{internal}$ = constant.
    $n*V_{chamber} \ll V_{basin} \implies$ flow continuity.
  - The internal pressure must be small compared to the atmospheric pressure, but higher than the used liquid saturation pressure:
    $P_{saturation} < P_{intérieur} < P_{atmosphérique}$
  - The monitoring system should be synchronized to get one cycle per second.

- Before vacuum settling inside the tank (pressure decrease) by vacuum pump, a limit level must be set to keep the liquid level constant by the means of the hydraulic pump.

III- Monocylindric atmospheric pressure machine (dimensioning example and power calculations)

1- *System study* (Tank + cylinder + piston)

Given the following system :

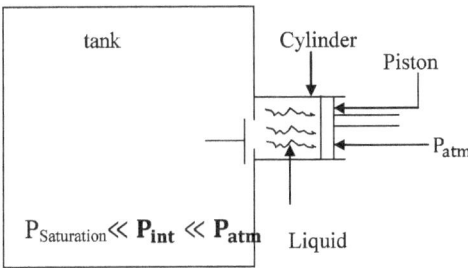

- The goal is to calculate the piston speed during its displacement from the initial location to the final location.
- It is needed to know Δt which is the duration of the piston displacement from its initial location to the final location. But to calculate the approximate piston speed, it is proposed to use Bernoulli theorem for the system under the following conditions:

* $P_{internal} = 20$ KPa
* $P_{atmosphéric} = 10^5$ Pa
* $S_A > S_B$ (S_A is slightly higher than S_B)
* Piston friction force is negligible
* Liquid friction force is negligible.

   S_A : Piston section

   S_B : Opening section

Bernoulli theorem results in the following:

$$\frac{\rho V_A^2}{2} + \rho g Z_1 + P_A = \frac{\rho V_B^2}{2} + \rho g Z_B + P_B$$

$$\Leftrightarrow \frac{1}{2g}(V_B^2 - V_A^2) = \frac{(P_A - P_B)}{\rho g}$$

$$\Leftrightarrow V_B^2 - V_A^2 = \frac{2(P_A - P_B)}{\rho}$$

Or on a $V_B S_B = V_A S_A \Rightarrow V_A = V_B \frac{S_B}{S_A}$

$$V_B^2 - (V_B \frac{S_B}{S_A})^2 = \frac{2(P_A - P_B)}{\rho}$$

$$V_B^2 \left(1 - (\frac{S_B}{S_A})^2\right) = \frac{2(P_A - P_B)}{\rho}$$

Avec $S_A > S_B \Rightarrow (\frac{S_B}{S_A})^2 < 1$

$$V_B^2 = \frac{2(P_A - P_B)}{\rho \left(1 - (\frac{S_B}{S_A})^2\right)}$$

$$V_B = \sqrt{\frac{2(P_A - P_B)}{\rho \left(1 - (\frac{S_B}{S_A})^2\right)}}$$

Et $V_A = \frac{V_B S_B}{S_A}$

2- *Dimensioning example and power calculation :*
a- *Piston power :*
- The used liquid is water, volumic mass $\rho = 1000 Kg/m^3$
- The piston diameter $d = 0.2$ m

- $F_p = (P_{atm} - P_{int}) \times S_{piston}$
  $AN: F = (10^5 - 20 \times 10^3) \times 3.14 \cdot 10^{-2} = 2512\ N$
- V_A =24.5 m/s but if we take into account the friction between the piston and the cylinder (steel over steel lubricated µ= 0.12) and the friction coefficient between water and steel is µ=0.065,
  As an estimate V_{Piston} = 20m/s , then P= $F_P * V_P$ = 20 * 2512 = 50240w
  = 50.240 Kw = 68.26 Cv

b- Power of the hydraulic pump :

- The hydraulic pump has to elevate water having a volumic mass of ρ= 1000Kg/m³
- The pump flow is $Q_v = n * V_{Chamber}/S$.
- The intermediate chamber is a parallelepiped rectangle wich has the dimensions a=25cm ; b=25cm ; l=40cm ; linked to a cylinder that has a diameter of 20cm and containing a piston that has the same diameter and a displacement stroke of 20 cm.

$V_{parallelepiped}$ = l*a*h = 0.025 m³ = 25L

$V_{cylinder} = \pi * \frac{d^2}{4} * c$ = 0.0314*0.2 = 0.00628 m³ = 6.28L

$V_{Chamber} = V_{parallelepiped} + V_{cylinder}$ = 0.03128 m³ = 31.28 L

Soit $V_{Chamber}$ = 32 L= 0.032 m³

It is needed to have a pump that has a flow of 32liters/s to elevate water quantity flowing into the tank every second:

*Calculation of the hydraulic pump power

The hydraulic pump elevates water from the tank where the internal pressure is $P_{interieur}$= 20Kpa to the basin which is under atmospheric pressure $P_{atmosphérique}$ = 10^5 Pa and which is at 10 m height h=10m

The pressure loss is estimated at 0,1 m

The used fluid is water which volumic mass is ρ = 1000 Kg/m^3

Let us apply the Bernoulli formula to the following system:

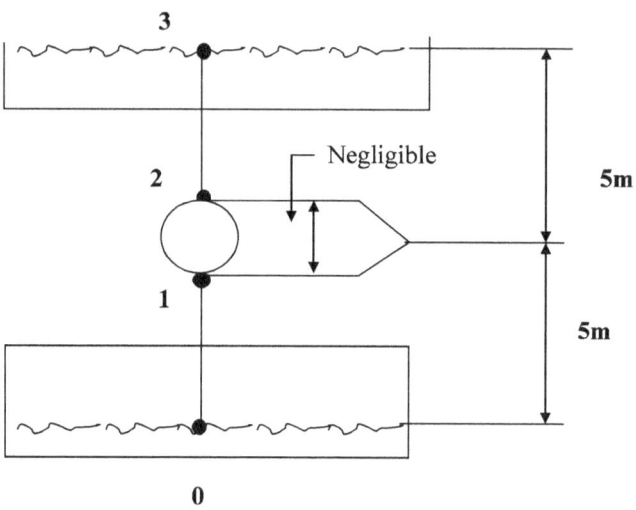

$$\frac{P}{Q_V} = P_3 - P_0 + \frac{1}{2}\rho(C_2^3 - C_0^2) + \rho g((Z_3 - Z_0) + J_{0-3})$$

$$P = Q_V[P_3 - P_0 + \frac{1}{2}\rho(C_2^3 - C_0^2) + \rho g((Z_3 - Z_0) + J_{0-3})]$$

$$P_3 = P_{atm} = 101{,}325 KPa = 10^5 \text{ Pa}$$

$$P_0 = 20 KPa = 20 \cdot 10^3 \text{ Pa}$$

$$C_3 = C_0 = 0 \text{ (still fluid outside the pipe)}$$

$C_2^3 = C_0^2$ (Discharge pipe diameter is equal to the suction pipe diameter, so the flow remains constant)

$$J_{0-3} = 0{,}1 \times 10 = 1m \text{ of fluid is added to } Z_3 - Z_0$$

It results in $P = Q_V\left[P_3 - P_0 + \frac{1}{2}\rho(C_2^3 - C_0^2) + \rho g((Z_3 - Z_0) + J_{0-3})\right]$
$$= 32 \times 10^{-3} \times [80 \times 10^3 + 10^3 \times 9.8(10 + 1)]$$
$$= 32 \times (80 + 107.8) = 6009.6W = 6.009 Kw$$

This pump is drived by an electric engine that has a global efficiency of 80%.

The consumed electric power is $P_{Consumed} = \frac{6}{0.8} = 7.5 \text{Kw}$

C- Power of the monitoring system

In the case of a monocylindric machine, the monitoring system consists of a compressor and two compressed air cylinders
- We need to calculate the force intensity that allows to displace the slide, then to calculate the monitoring system power
- Given the following system

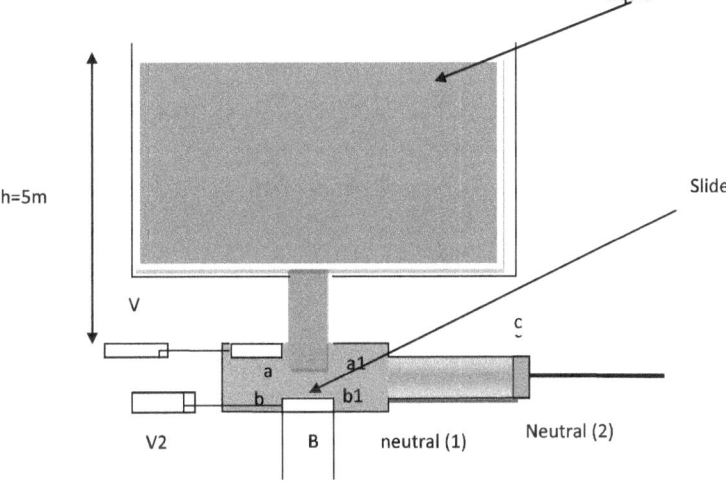

The goal is to calculate the force intensity that allows to displace the slide.

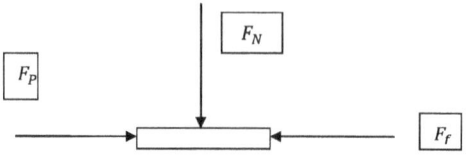

- The slide mass m=1Kg
- The slide diameter d=0.2
- $F_N = (P_{atm} + \rho gh) \times S + m \times g$
- Friction coefficient $\mu = 0.12$ (*steel over steel on lubricated faces*)

Then we have :

$$F_N = (P_{atm} + \rho gh) \times S + m \times g \quad \text{AVEC} \begin{cases} \rho = 1000 Kg/m^3 \\ h = 5m \ (eau) \\ P_{atm} = 10^5 Pa \end{cases}$$

$F_N = (10^5 + 10^3 \times 9.8 \times 5) \times 0.0314 + 1 \times 9.8 = 4688.4 N$

↔ $F_{Thrust} = F_f = \mu F_N = 0.12 F_N = 562.6$

- To displace the slide it is necessary that $F_{Poussée} \geq F_f$

The cylinder is selected at thrust $F_{Thrust} = 1000$ N to assure a good cylinder efficiency, and we calculate the required cylinder pressure to determine the thrust.

- A double effect cylinder is chosen, diameter is D=25mm and c=200m

with $\begin{cases} D: cylinder\ section\ diameter \\ V_{cylinder}: cylinder\ volume \\ C: displacement\ stroke\ of\ the\ cylinder\ piston \end{cases}$

$S = \frac{0.025^2}{4} \times \pi = 4.9 \times 10^{-4} m^2$

$V_{vérin} = S \times C = 4.9 \times 10^{-4} \times 0.2$
$= 9.8 \times 10^{-5} m^3 = 0.098 \times 10^{-3} m^3$
$= 0.098 L$

Then $F_P = P \times S \implies P = \frac{F_P}{S} = \frac{1000}{4.9 \times 10^{-4}} = 20.4$ bar

Note: Compressed air cylinder setting is done by modifying the air flow. This is accomplished by modifying the nominal compressed air port in the appropriate distributor.

* Compressor selection

To select the right compressor, one has to know the sum of compressed air consumptions of all accessories simultaneously linked to the compressor. The sum should be multiplied 1.5.

This result is equivalent to air flow as liters per minute needed to operate all the compressed air accessories.

In the current system, the double effect compressed air cylinder makes a cycle per second et it consumes 0.196 L/s because $V_{cylinder} \times 2 = 0.098 \times 2 = 0.196$ L/S

The cylinder consumption per minute is $0.196 \times 60 = 11.76$ L/mn.

A compressor engine rated between 0.2 Cv and 5.5 Cv generates a pressure between 1.5 bar and 500 bar and gives an air flow between 10L/mn and 500 L/mn.

In the case of the current required cylinder, a 500 liters/mn compressor rated 5.5Cv can drive 28 cylinders that have the same dimensions already selected which corresponds to 14 pistons in a multicylindric atmospheric pressure machine.

=>→ 28×cylinder consumption/mn×1.5=28×11.76×1.5

$$=493.5 \text{L/mn} \leq 500 \text{L/mn}$$

2. Monocylindric machine rated 1 MW example

The piston diameter is selected at $D=1m$ then $F=P*S=62800N$. As V. piston $=20$m/s then Power$=1256$KW.
It is supposed that water volume suctioned during every cycle is $V=1m^3$, then the needed flow is $Qv=1m^3/s$. To elevate such a volume to the basin, a propeller pump type is selected to have a flow $180m^3/h$ (50 l/s). As a result, 20 propeller pumps are needed to get a flow of Qv=1m3/s. Every pump has to elevate water to the basin which is at 4 m height. One pump needs an electric power P-electric = P=3.5Kw, then the consumed power is Pt=70Kw.

To accomplish both slides displacement, a 5,5Kw compressor is needed.

A power loss of 180,5Kw can be considered although the theoretical piston speed is decreased by 4,5m/s to compensate the friction effect. Then remaining power is 1MW.

The atmospheric pressure machine has a very high conversion rate whatever is location and climatic conditions. And if we assume in the most defavorable cases annual maintenance duration of two months we can get a production capacity of 7200MWh/year from 1MW installed atmospheric power.

IV- Multicylinders atmospheric pressure machine

a-Operation

The multicylinders atmospheric pressure machine has several pistons linked to a crank shaft or to a crank shaft equipped with flywheel. It has the same operating principle than the monocylindric machine.
There are two ways to convert translation movement to rotating movement.
* First method: Conversion through a crank shaft with a flywheel.
This method consists of linking a flywheel to the crank shaft. Its role is to store energy during the suction phase (driving stroke) to give it back during the intake phase and the synchronization phase.
* Second method: conversion through crank shaft only.
This method consists of multiplying cylinders providing that their respective driving phases (suction stroke) are well distributed during the synchronization phase (t=1s) and the intake phase.

b-Dimensioning example of the multicylinders machine

Atmospheric pressure machine with 28
- Piston diameter d=20 cm
- Piston stroke c=20 cm
- Parallelepiped dimensions a=25 cm, b=25 cm, l=40 cm
- Used liquid : water, volumic mass $\rho = 1000 \, Kg/m^3$
- 28 hydraulic pumps (the flow of each pump is Q_v=32 L/s)

- Two compressors (each compressor is rated 5.5Cv, its flow is 500L/mn and a maximal pressure 500 bar)

As a result :

* P_{piston}=50.240 KW
* $P_{total\ pistons}$=28×50.240=1406.72 KW
* P_{pump}=7.5 KW
* $P_{total\ pumps}$=28×7.5=210 KW
* $P_{copressor}$=4.048 KW
* $P_{total\ copresors}$=2×4.048=8.096 KW
* $P_{machine} = P_{total\ pistons} - (P_{total\ pumps} + P_{total\ compressors})$

   = 1406.72 − (210+8.096)

   =1188.624KW

   =1.188MW

Note : the same machine power can be obtained by increasing the pistons diameter while decreasing the pistons number.

V- Atmospheric micro-power plant

- The atmospheric micro-power plant is a small scale atmospheric power plant producing a small energy amount.

- The atmospheric micro-power plant has few atmospheric pressure machines, or a well determined number of pistons linked to one atmopheric system to generate electricity to factories, homes, hospitals, ,...etc.

VI- Atmospheric power plant

The multicylinders atmospheric pressure machine can be used to build a large scale power plant to produce electricity.

-EXAMPLE

900 Machines rated 1MW each have a capacity of 900MW which is equivalent to one nuclear reactor.

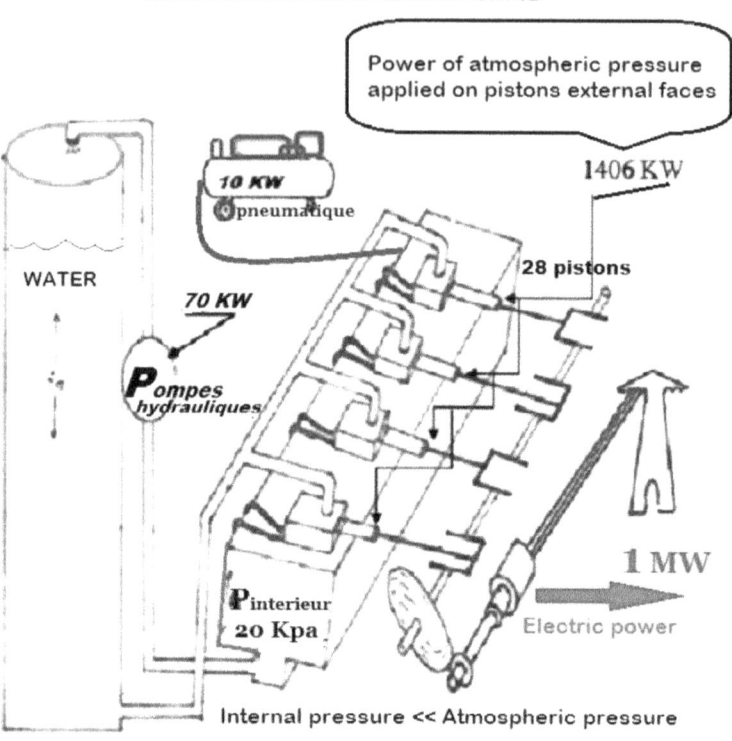

INVENTION OF

A NEW TECHNOLOGY TO PRODUCE RENEWABLE ENERGY BY THE MEANS OF WATER HYDROSTATIC PRESSURE

Hydrostatic machine

Copyright Sadok JABLI, 2015

Abstract

This invention is related to a hydrostatic machine submerged in water at a determined depth (10m, 20m, or deeper) which is used to produce renewable electric energy from the energy transferred by the action of hydrostatic pressure on the piston.

It is a new technology to produce renewable energy at a very high conversion rate, totally independent from location and climate, at a very low manufacturing cost compared to other renewable energy sources.

The document includes all required scientific and technical information

Hydrostatic machine

This invention is related to an hydrostatic machine immersed in water at a determined depth (10m, 20m, or deeper) which is used to produce renewable electric energy from the energy transferred by the action of hydrostatic pressure on the piston.

Figure (1)

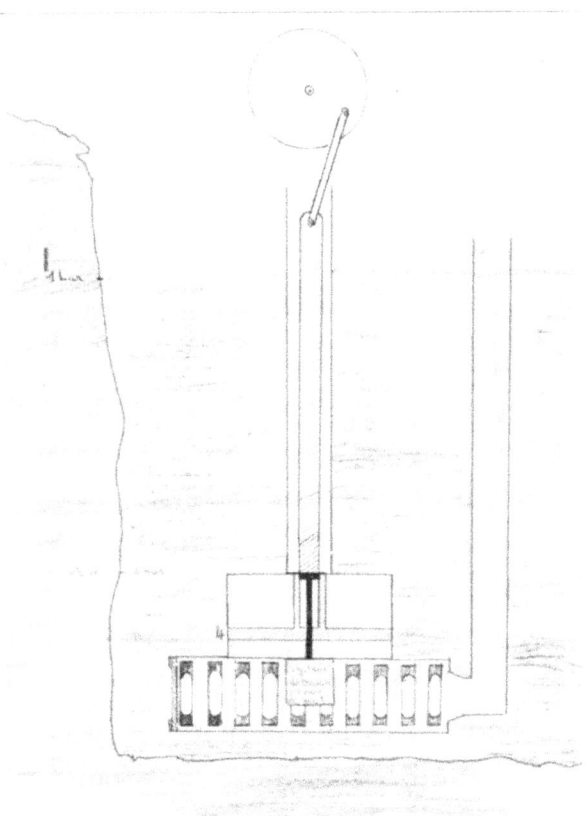

The hydrostatic machine components are:

- A jacket equipped with a piston linked to a rod and crank system that activates an electric generator. It is built in such a way that the piston upper face is always under atmospheric pressure and the lower face (contact area) is communicating with water at hydrostatic pressure only once per cycle.
- A cylinder located under the jacket, having two openings « hydrostatic openings » and delivery pipes which are open to the external environment at atmospheric pressure.
- The hydrostatic openings area is half the piston cross section area. They allow the water to flow into the cylinder at hydrostatic
  $P = P_{atmosphéric} + \rho g h$
- Delivery pipes are mounted next to the hydrostatic openings on the same plane in such a way that beside every hydrostatic opening there is a delivery pipe communicating with the external environment where pressure $P = P_{atmospheric}$
- Delivery pipes cross section area is the same than the hydrostatic openings ones.
- A sliding cylinder activated by a monitoring system has a function to open and close the delivery pipes and the hydrostatic openings in order to put the piston at the differential pressure between the hydrostatic pressure and the atmospheric pressure

  $\Delta P = P_{hydrostatic} - P_{atmospheric}$ and the piston is released at every cycle.

- The cylindrical slide control is accomplished through a monitoring system that allows the slide rotation in both directions in order to change the prevailing conditions (open, closed) of the delivery pipes and the hydrostatic openings at every cycle.

- The monitoring system consists of a cylinder, position sensors mounted to the (jack + piston) and a compressor.

- The delivery pipes and the cylinder hydrostatic openings are separated by space distances. Every space distance has the same distance than the opening that exists in the sliding cylinder, which allows the system to fulfill all required conditions (open, closed) before the next condition occurs.

- Figure (2)

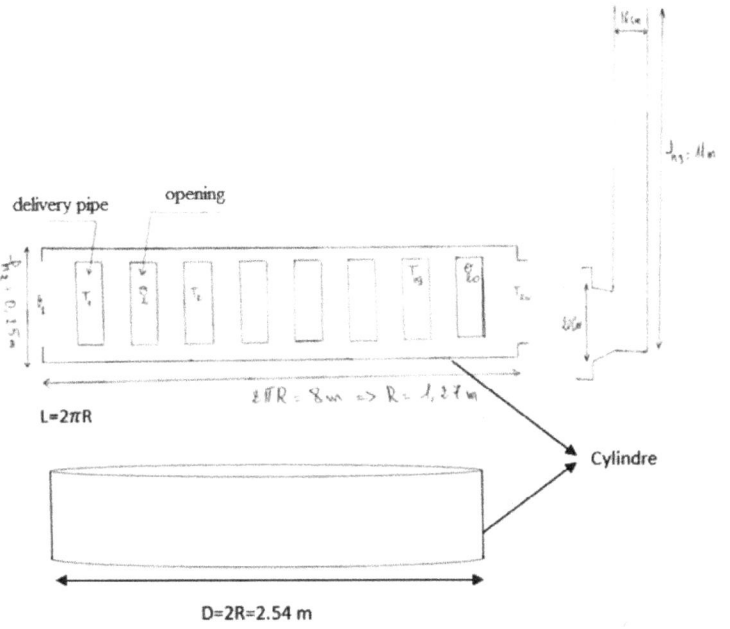

L=2πR

D=2R=2.54 m

Cylindre

Successive conditions are as follows:

1- $\begin{cases} hydrostatic\ openings: closed \\ delivery\ pipes : closed \end{cases}$

2- $\begin{cases} hydrostatic\ openings: open \\ delivery\ pipes: closed \end{cases}$

3- $\begin{cases} hydrostatic\ openings: closed \\ delivery\ pipes: closed \end{cases}$

4- $\begin{cases} hydrostatic\ openings: closed \\ delivery\ pipes: open \end{cases}$

- The operating conditions mentioned above are the 4 stroke cycle of the machine that is continuously reproduced during operation.

HYDROSTATIC MACHINE OPERATING PRINCIPLE
Phase 1

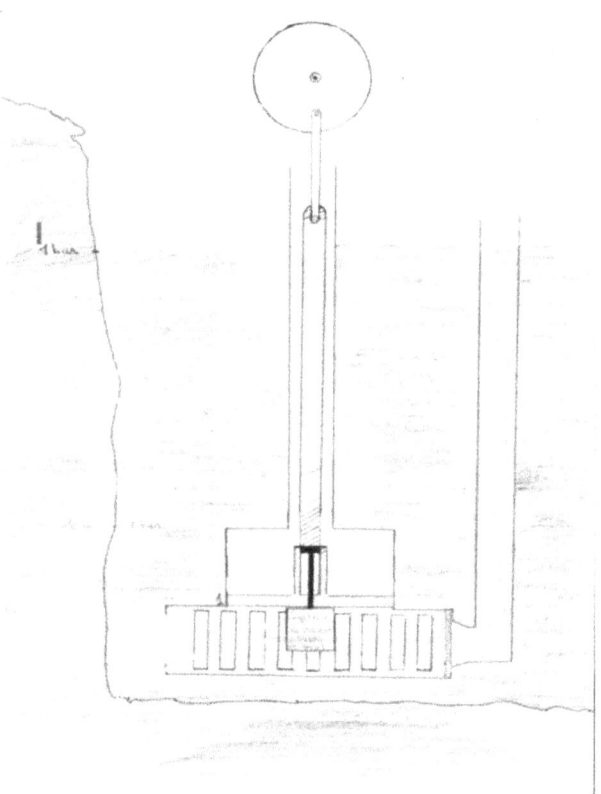

Figure (2)

- The piston is at point (1)
- The hydrostatic openings are opened and the delivery pipes are closed
- The water at pressure $P = P_{\text{atmosphéric}} + \rho g h$ flows into the cylinder.
- The piston moves up due to the pressure difference:
  $\Delta P = P_{\text{hydrostatic}} - P_{\text{atmospheric}}$ creating a natural push force $F = \Delta P \times S$

- Where
$\begin{cases} S: \text{piston cross section area} \\ \Delta P = \text{Hydrostatic pressure } - \text{ Atmospheric pressure} \end{cases}$

Phase 2

Figure (3)

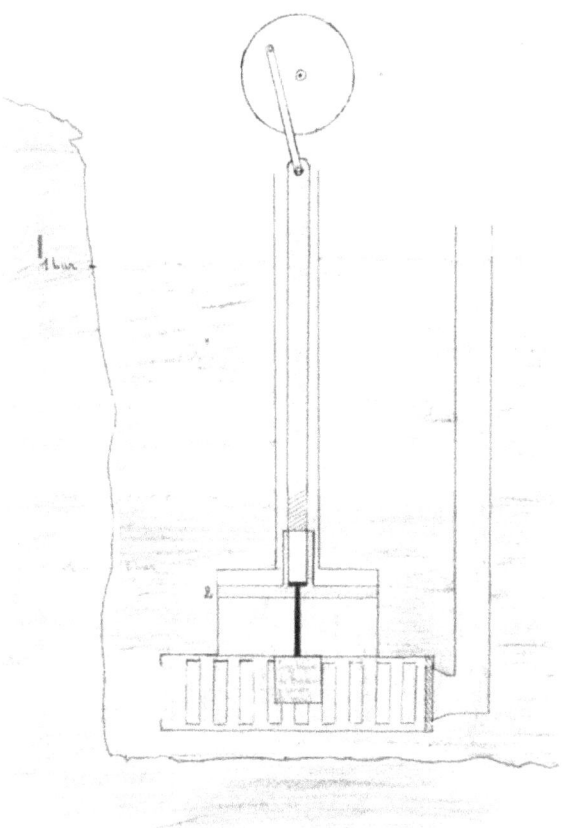

The piston reaches point (2) and a sensor « C » activates the sliding cylinder to rotate

- At the end of the piston stroke at point (3) the sliding cylinder closes the hydrostatic openings and opens the delivery pipes which are communicating with the external environment at pressure P= $P_{atmospheric}$.

Phase 3

Figure (4)

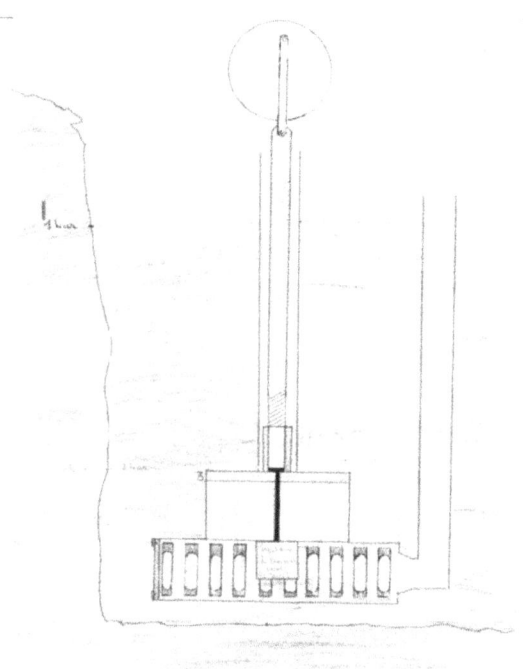

The piston moves downward due to its weight $P_{piston} = m_{piston} \times g$ providing that the piston weight must be slightly greater than the force required to push the remaining water in the system whenever the hydrostatic openings are closed.

Piston weight ≥ $F_{push\ force\ to\ evacuate\ remaining\ water}$

Phase 4

Figure (5)

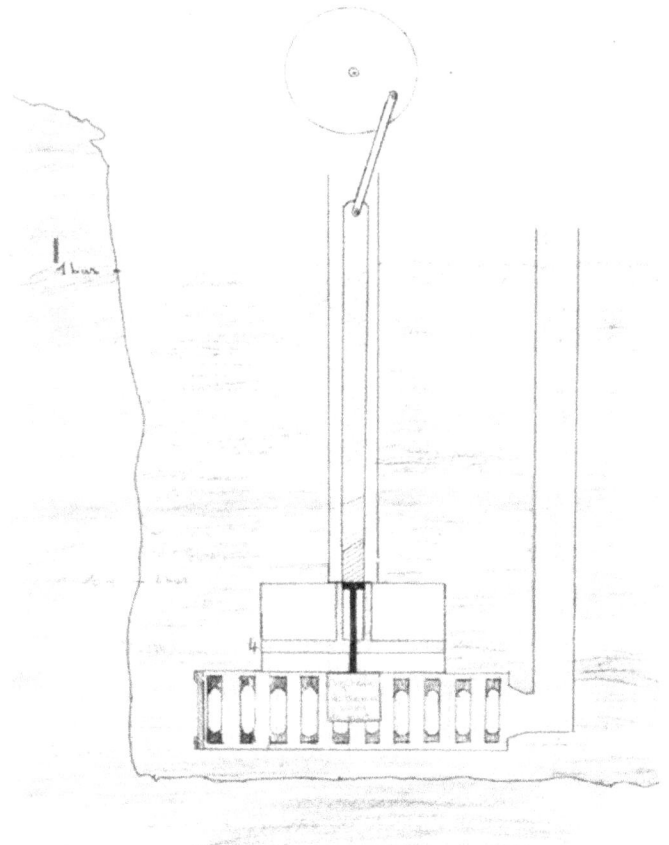

- When moving downward, the piston reaches the point (4) and a sensor « C1 » activates the sliding cylinder to rotate in the opposite direction
- At point (1) the sliding cylinder closes the delivery pipes and opens the hydrostatic openings, and the cycle restarts.

Dimensions selection and machine power calculation
Figure (6)

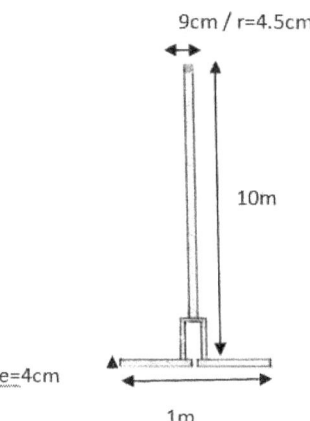

Thickness e = 4cm = 0.04m

The used material is duplex steel, density $\rho = 7850 \, kg/m^3$

Piston diameter d=1m

Piston cross section area S=0.785 m²

Piston radius r =4.5cm=0.045m

Piston height (thickness) e= 4cm =0.04m

Depth H=10m

<u>Piston mass :</u>

$m_{piston} = \pi R^2 h\rho + \pi r^2 e\rho$

$= 246.49 + 499.1 = 745.5 \, kg$

Piston weight :

$P_{piston} = m_{piston} \times g = 7306.3$ N.

Piston speed calculation

The following system is considered:

Figure (7)

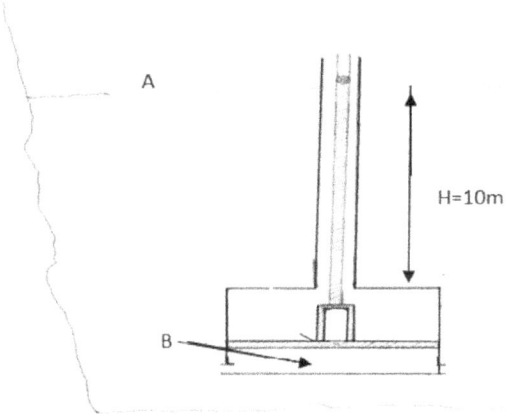

Assumed that:

The piston moves without friction

Hydrostatic openings total area is equal to the half piston cross

$S_{Total\ ouvertures} = \frac{1}{2} S_{piston} = 0.392$ m²

Bernoulli theorem applies between point A and point B

$$\frac{\rho V_A^2}{2} + \rho g Z_A + P_A = \frac{\rho V_B^2}{2} + \rho g Z_B + P_B$$

where $P_A = P_B = P_{atmosphérique.} = 1 \text{bar}$

Water flowing speed in the cylinder is $V_1 = \sqrt{2gh} = 14$ m/s

The relationship between water flowing speed V_1 and the piston speed V_2 (without friction) is as follows :

$V_1 S_1 = V_2 S_2 \Rightarrow V_2 = V_1 \frac{S1}{S2} = \frac{1}{2} V_1 = 7$ m/s

But if friction is considered between the piston and the jacket (steel over steel, lubricated contact) $\mu = 0.12$ and friction between water and steel $\mu = 0.065$, the piston speed is then estimated as $V_{piston} = 4.5$ m/s.

The piston driving force calculation for the driving phase where piston moves upward along a distance of (total depth) H=10m

- The force acting on the piston due to hydrostatic pressure at a depth H=10m is: $F_{hydrostatic} = P_{hydro} - P_{atm} = \Delta P \times S_{piston} = 78500$ N

  Avec $\begin{cases} \text{Hydrostatic pressure } = 2\text{bar} \\ \text{Atmospheric pressure } = 1\text{bar} \end{cases}$

- The resistant force is the sum of the piston weight strength and the needed force to rotate the sliding cylinder.
- The piston weight is $P_{piston} = m_{piston} \times g = 7306.3$ N
- The needed force to move the sliding cylinder $F_{p.slide}$ is calculated as follows :

  There are 40 rectangular openings in the sliding cylinder. Every one of them has a width of l=10cm and a length of L=20cm.

  Between every two neighboring openings there is a rectangular separating space that is 10 cm wide and 20 cm long.

  To get these 40 openings separated by 40 spaces that have the same area, the needed sliding cylinder must have a radius R=1.27m, a height h=25cm and a thickness e=3cm.

  Sliding cylinder mass

  $m = \pi(R^2 - r^2)h\rho = \pi((1.27)^2 - (1.24)^2) \times 0.25 \times 7850$

  $m_{slide} = 464$ kg

  The sliding cylinder weight is $P_{slide} = m_{slide} \times g = 7306.3$ N

  Since $F_{p.slide} = F_{friction} = \mu F_N = \mu P_{slide} = 0.12 \times 4547.3 = 545.6$ N

  To move the sliding cylinder, we must have $F_{p.slide} \geq F_{ffriction}$

  We assume that $F_{p.slide} = 1000$N in the following calculations

(An air compressor whose engine is rated 1.5 KW can deploy this force.)

Thus $F_{motor} = F_{hydrostatic} - F_{resistant}$

$\qquad = F_{hydrostatic} - (P_{piston} + F_{p.slide})$

$\qquad = 78500 - (7306.3 + 1000) \cong 70000 N$

$F_{motor} = 70000N$ is the gross natural force recovered by the hydrostatic machine for the previously chosen dimensions.

But in order to restart the described cycle and to make the machine running correctly, the piston must have the capability to move downward again.

That is why the piston weight must be slightly greater than the needed force to push the amount of water remaining in the system when the hydrostatic openings are closed and the delivery pipes are opened.

Calculation of the needed force to push the remaining water in the system:

Figure (8) fh9

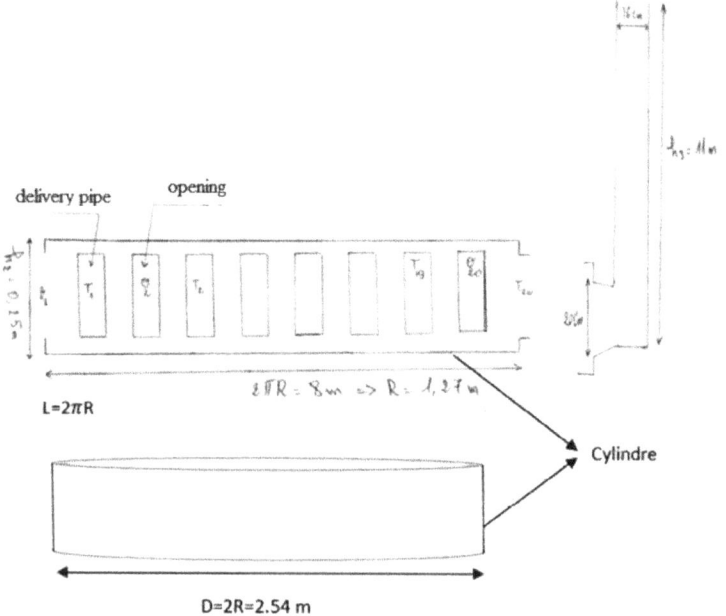

L=2πR

D=2R=2.54 m

$V_{water} = V_{cylinder} + V_{jacket} + V_{pipes}$

- $V_{jacket} = \pi R^2 h_1$ avec $\begin{cases} h_1 = 0.5 \\ R = 0.5 \end{cases}$

     $= 3.14 \times (0.5)^2 \times 0.5 = 0.392 \text{ m}^3$

- $V_{cylinder} = \pi (R_1)^2 \times h_2 = 3.14 \, (1.27)^2 \, 0.25 = 1.266 \text{ m}^3$
- $V_{pipe} = \pi (R_2)^2 \, h_3 = (3.14)(0.08)^2 \times 11 = 0.221 \text{ m}^3$
- $V_{pipes} = 20 \times V_{pipe} = 20 \times 0.221 = 4.421 \text{ m}^3$

Thus, the volume of the remaining water in the system when hydrostatic openings are closed and delivery pipes are opened is :

$V_{water} = V_{cylinder} + V_{jacket} + V_{pipes}$

       $= 1.266 + 0.392 + 4.421$

$= 6.079 \text{ m}^3$

The mass of the remaining water in the system is :

$M_{water} = V_{water} \times \rho = 6291.7 \text{ kg}$ where $\rho = 1035 \text{kg/m}^3$ is the sea water density.

Mass of the remaining water in the system

$P_{water} = m_{water} \times g = 61659.2 \text{N}$

The needed force to push the remaining water

$F_{push} = F_{friction} = \mu \times F_N = \mu \times P_{water} = 0.065 \times 61659.2 = 4007.8 \text{ N}$

where $\mu = 0.065$ is the friction facor between water and duplex steel.

We chose $F_{push} = 5000\text{N}$

Then

$P_{piston} = 7306.3 \text{ N}$; $F_{slide} = 1000\text{N}$; $F_{push} = 5000\text{N}$

$\Rightarrow P_{piston} > F_{slide} + F_{push}$

CALCULATION OF THE MACHINE POWER

We have $F_{motor} = 70000\text{N}$

Then, the piston mechanical power is :

$P_{mec\text{-}Piston} = F_{motor} \times V_{piston} = 315 \text{ KW}$

The machine electrical power is :

$P_{\text{électric machine}} = P_{mec\text{-}Piston} \times 0.8$

$\qquad = 252 \text{ KW}$

FRIGORIFIC CYCLE ATMOSPHERIC MACHINE

INVENTION

Frigorific cycle atmospheric machine

BACKGROUND

This invention consists of an atmospheric machine using frigorific cycle to produce renewable electric energy out of atmospheric pressure energy coming from nature and applied on the piston(s) external face.

For this machine, ambient air pressure is the origin of the system driving force.

The process consists of installing a pressurized empty cylinder into a frigorific cycle without disturbing its functioning operation. This is done in such a way that frigorigenic evaporates inside the cylinder and pushes the piston while it condensates to create vacum again which provoques piston displacement because of atmospheric pressure applied on its external face. And the cycle restarts again.

The piston is linked to a flywheel that accumulates atmospheric pressure energy as cinetic energy during the piston driving phase and to restore it during a period « t » required for filling the cylinder with frigorific liquid and for condensation, without any important change in the rotating speed.

OPERATION PRINCIPLE

Image 1

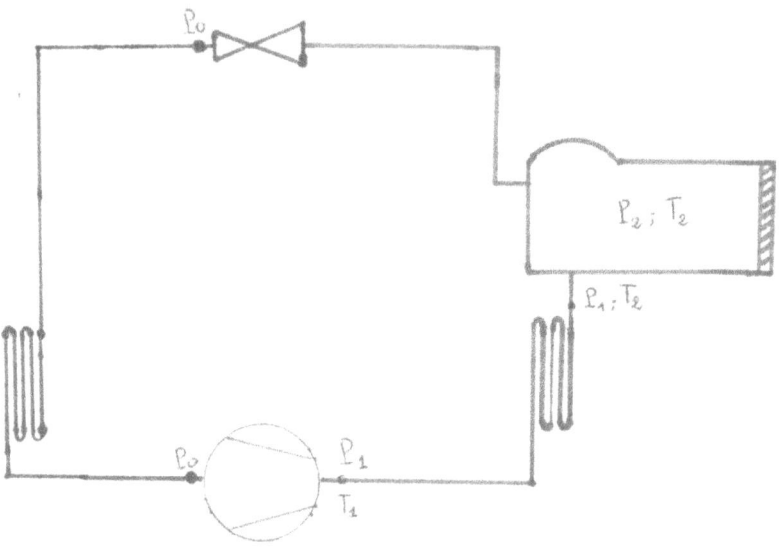

The frigorific liquid is moving in a closed circuit and making 5 stages cycles:

1-Compression:

While entering the compressor the frigorific liquid is a steam at pressure P_0.

At the compressor outlet, the frigorific fluid is a steam at pressure P_1 and at high temperature T_1.

2-Condensation:

At condenser entrance the frigorific fluid is a steam.

While going through the condenser the fluid (which is at a high temperature), loses its thermal and becomes a liquid at the same input pressure P_1 and at at temperature $T_2 < T_1$

3- Primary expansion :

Image 2

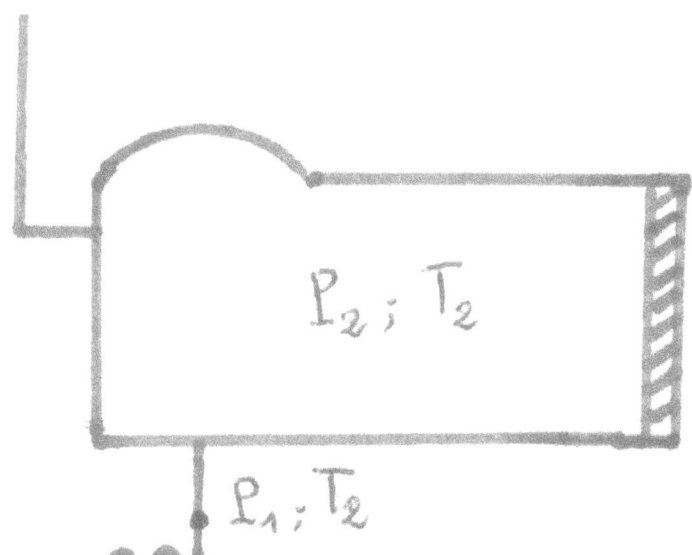

The primary expansion is accomplished by the means of a regulation chamber linked to a cylinder that is equipped with a piston.

The regulation chamber includes an expansion spring, a membrane connected to a valve and a braking rod monitored by a push rod.

This system allows to close the frigorific fluid (liquid state) inlet port, as soon as the cylinder internal pressure equals the required pressure P_2, and to open it after steam condensation.

The pressure P_2 is the frigorific liquid evaporation pressure at temperature T_2 of the fluid at the condenser outlet.

Inside the cylinder, the liquid evaporates at temperature and pressure (T_2, P_2), at this level cylinder external temperature is very low compared to T_2 because of a fan assembled to the evaporator. It blows cold air outside the cylinder. Thus, the frigorific fluid (steam state) loses its thermal energy and liquefies.

This prompt condensation creates vacuum again in the cylinder and the piston moves under the effect of atmospheric pressure until the end of its stroke allowing the opening of the inlet and the flow of the frigorific liquid entering the cylinder.

During the required time to reach P_2 the piston makes a free movement back and forth due to a flywheel that stores energy during the driving phase and restores it back during the period « t »

The cylinder outlet orifice is always open.

At the exit of the bloc «regulation chamber-Cylinder» the frigorific fluid is in the liquid state at pressure P_2.

4-Secondary relaxation:

The fluid passes through the regulator, its pressure as well as its temperature decreases.

The regulator allows also the flow regulation of the fluid along the closed circuit.

At the regulator exit the fluid is at pressure P_0.

5-EVAPORATION:

While going through the regulator, the frigorific fluid (at low temperature) catches thermal energy, and that is why it evaporates.

General scheme of the Frigorific cycle atmospheric machine

Image 2

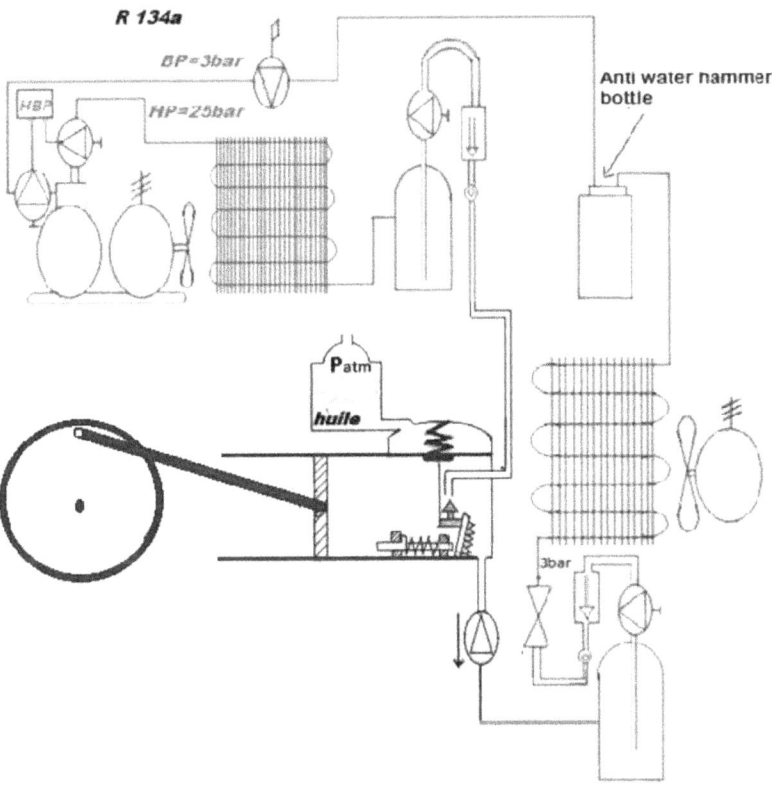

Image 3

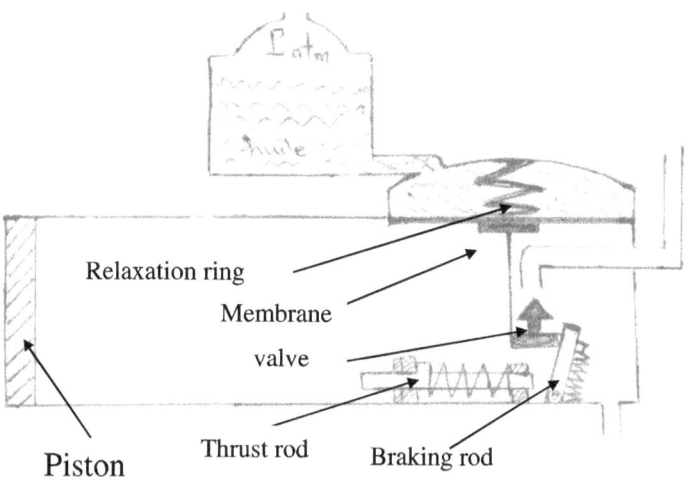

The bloc « Regulation chamber-cylinder » consists of a regulation chamber connected to cylinder that has a piston.

The regulation chamber includes a relaxation ring, a membrane linked to a valve and a braking rod activated by a thrust rod.

At the condenser exit, the frigorific fluid is in a liquid state, at high pressure and at the T_2 temperature.

Inside the cylinder, the valve closes the orifice as soon as the pressure reaches the value of P_2 pressure which corresponds to the liquid evaporation pressure at temperature T_2.

The liquid evaporates inside the cylinder at point (P_2, T_2).

The cylinder external temperature is very low compared to the T_2 temperature due to the fan mounted on the evaporator that blows cold air on the outer wall of the cylinder. As a consequence, the frigorific (steam) its thermal energy and becomes a liquid again.

The steam prompt condensation creates vacuum into the cylinder.

-The force F having the value F=P.S pushes the piston until the stroke end,

-the piston pushes the thrust rod,

-the thrust rod pushes the braking rod,

-The liquid inlet orifice opens and the cycle restarts again,

The liquid outlet orifice is always open.

The « regulation chamber-cylinder » bloc dimensions depend on the volumic flow of frigorific fluid at the condenser exit.

Dimensions are determined as follows:

Inside the cylinder the fluid undergoes a transformation supposed to be isothermal,

then $P_1 \times V_1 = P_2 \times V_2$

Where P_1 : fluid pressure at the condenser exit.

P_2 : Fluid pressure inside the cylinder,

V_1 : Volume of fluid flowing during the period of time « t » needed to reach the pressure P_2 inside the « regulation chamber-cylinder » bloc,

V_2 : Volume of the bloc « regulation chamber-cylinder »

The frigorific fluid volume flowing during the period of time « t » corresponds to « N » times the volumic flow of the frigorific fluid at the condenser exit.

N= t/s with $\begin{cases} t = flowing\ time\ needed\ to\ reach\ P_2 \\ s = \quad one\ seconde \end{cases}$

This is determined taking into account the volume of « regulation chamber-cylinder » bloc and the section of the piston used to produce energy.

To size the system, volumic flow blown by the compressor is used as a starting point.

The method is the following:

1-Calculation of hourly volume at suction:

$$Q_{Va} = \frac{Q_{Vb}}{\eta V_0} \qquad \text{Avec} \quad \eta V_0 = 1 - 0.05 \times P_1/P_0$$

and $\begin{cases} P_1 = \text{Pressure at compressor exit} \\ P_0 = \text{Pressure at compressor inlet} \end{cases}$

Q_{Va} : Hourly volume at compressor suction

2-Fluid massic flow calculation

$$Q_m = \frac{Q_{Va}}{V'_1}$$

Q_m : Fluid massic flow in the circuit

V'_1 : Compressor massic volume at steam phase into the evaporator (at point 1)

3- Fluid volumic flow at compressor outlet

$$Q_V = \frac{Q_m}{\rho}$$

ρ : Frigorific liquid volumic mass at compressor (liquid phase, 45°C)

Calculation of the period of time « t » needed to reach P_2

$V_1 / V = N$ Then $t = N \times s$

With $\begin{cases} V = \text{Volume flowed during one second at condenser outlet} \\ V_1 = \text{volume flowed during t} \\ s = \text{second} \end{cases}$

The piston accomplishes only one driving stroke during the period of time « t »

Starting from this result the suitable flywheel is selected.

Dimension example and power calculation

a)-Calculation of piston speed

Let us assume that the piston moves without friction and that its speed is similar to the one of flowing water into a vacuum volume whose internal pressure is $P_{internal} = 20$ KPa.

Suppose the following system :

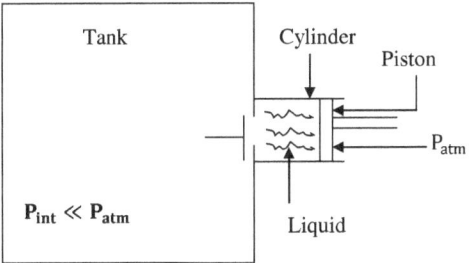

This system is under the following conditions:

- P_{int} of the tank is very low compared to atmospheric pressure
- Fluid saturation pressure corresponding to the temperature is lower than tank internal pressure, making that fluid remains in liquid state. .

$S_A > S_B$ et $f = 0$

S_A : Piston section

f : Coefficient of friction

S_B : Section of the opening

By applying the Bernoulli theorem:

$$\frac{\rho V_A^2}{2} + \rho g Z_1 + P_A = \frac{\rho V_B^2}{2} + \rho g Z_B + P_B$$

$$\Leftrightarrow \frac{1}{2g}(V_B^2 - V_A^2) = \frac{(P_A - P_B)}{\rho g}$$

$$\Leftrightarrow V_B^2 - V_A^2 = \frac{2(P_A - P_B)}{\rho}$$

Since we have $V_B S_B = V_A S_A \Rightarrow V_A = V_B \frac{S_B}{S_A}$

$$V_B^2 - (V_B \frac{S_B}{S_A})^2 = \frac{2(P_A - P_B)}{\rho}$$

$$V_B^2 \left(1 - (\frac{S_B}{S_A})^2\right) = \frac{2(P_A - P_B)}{\rho}$$

With $S_A > S_B \Rightarrow (\frac{S_B}{S_A})^2 < 1$

$$V_B^2 = \frac{2(P_A - P_B)}{\rho\left(1 - (\frac{S_B}{S_A})^2\right)}$$

$$V_B = \sqrt{\frac{2(P_A - P_B)}{\rho\left(1 - (\frac{S_B}{S_A})^2\right)}}$$

Then $V_{piston} = V_A = V_B \times \frac{S_B}{S_A}$

Thus for a piston having a diameter d=1m and an opening diameter d'=0.98m
We have $S_A = 0.785 m^2$ et $S_B = 0.753 m^2$

$$V_B = \sqrt{\frac{2(P_A - P_B)}{\rho\left(1 - (\frac{S_B}{S_A})^2\right)}} = 40 \text{ m/s}$$

$$\Rightarrow V_A = V_B \times \frac{S_B}{S_A} = 36 \text{ m/s}$$

If friction the piston and the cylinder is considered (steel over steel on lubricated surface μ=0.12m/s and steam condensation speed, the maximum real speed of the psiton during driving phase is assumed at V_{piston} =20m/s for a piston that has a diameter d=1m and a section of $S=0.785m^2$

Machine electric power calculation

The frigorific fluid used is R134a (it has no greenhouse effect on the ozone layer).

We prefer to use:

- A frigorific compressor with an optimized piston using R134a, and an operating volume of $V_b = 535 m^3/h$ consuming an electric power of 170 KW

-a fan consuming 8 KW

-a piston having a diameter d= 1m corresponding to $S=0.785\ m^2$

-the height of the cylinder and the height of the regulation chamber is h= 0.25m

Then $V_2 = 0.25 \times 0.785$

$\qquad = 0.196\ m^3$

For a cycle of R134a in between 3bar and 25bar the fluid liquefies into the condenser in between 30 °C and 50°C.

We chose a pressure of $P_2 = 9$ bar; corresponding to the R134a evaporation pressure at $T_2 = 40°C$.

Then we have $P_1 \times V_1 = P_2 \times V_2 \Rightarrow V_1 = \frac{P_2 \times V_2}{P_1}$

N.A: $V_1 = \frac{9 \times 0.196}{25} = 0.070\ m^3$

V_1 Corresponds to the frigorific fluid flowing during the period of time « t » needed to reach the pressure P_2 inside the bloc « regulation chamber-cylinder »

Then we have

$Q_{Va} = \frac{Q_{Vb}}{\eta V_0} = \frac{535}{0.58} = 922.4\ m^3/h$

$Q_m = \frac{Q_{vb}}{V'_1} = \frac{922.4}{0.05} = 1590$ Kg/h $= 5.12$ Kg/s

$\Rightarrow Q_V = \frac{Q_m}{\rho} = 0.004 \; m^3/s$

$\Rightarrow V = 0.004 m^3$ (Frigorific liquid volume passed during one second at condenser exit)

Then $N = \frac{V_1}{V} = 17.5 \Rightarrow t = 17.5s$.

Assuming « t » equals 20 seconds, and then the piston accomplishes a driving stroke every 20 seconds, and then moves freely during the remaining time « t ».

We select a flywheel giving the constant rotation speed similar to the average constant speed of the piston.

We assume that the piston average speed is 10 m/s.

We have **F= $P_{atmospheric} \times$ S =78500 N**

$\Rightarrow P = F \times V = 785\,000 W = 785$ KW $= 0.785$ MW.

The conversion efficiency rate of mechanical energy into electrical energy is 0.8
Then $P_{electrical} = 628$ KW $= 0.628$ MW.

Principle of energy conservation

Input energy = atmospheric pressure energy (converted into kinetic energy by the means of flywheel) + Energy consumed by the compressor + energy consumed by the two fans.

Output energy = electrical energy given by the generator.

Yield $\tau = 0.7$

Frigorific cycle atmospheric machine

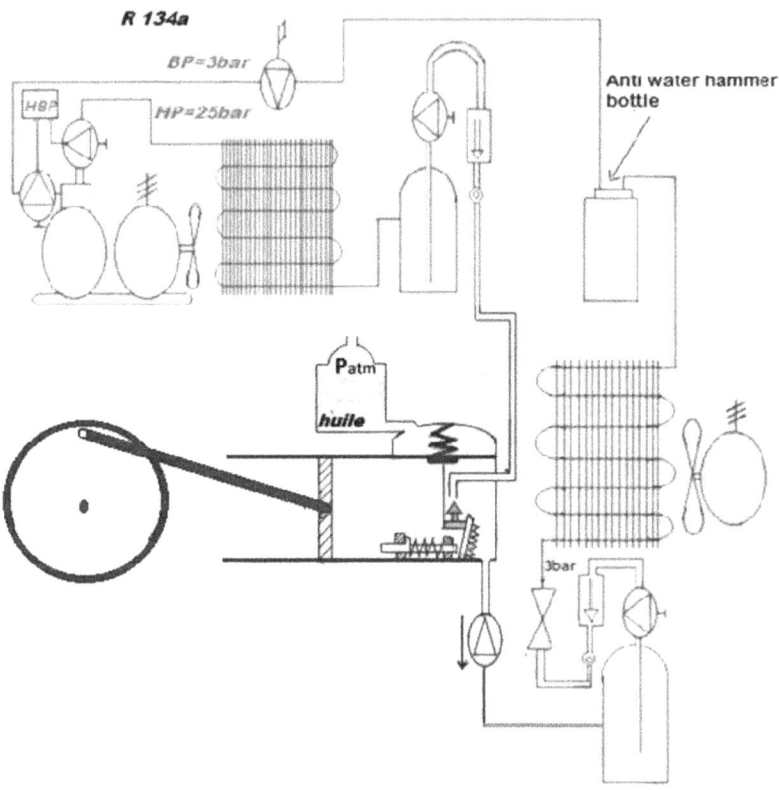

INVENTION OF A NEW PROTECTIVE DEVICE

INVENTION OF A NEW PROTECTIVE DEVICE

The protective device against indirect contact with metallic body without grounding is suitable to minimize risks resulting from failure of differential circuit breakers and from electric neutral regime malfunction. Thus, it protects millions of users all over the world who don't have grounding. Actually, the majority of electric injuries occur where grounding is missing.

Copyright Sadok JABLI, 2013

DESCRIPTION

I-PRESENTATION

This invention consists in a protective device against electric risks caused by leaking current flowing within the human body by indirect contact (metallic object accidentally energized without grounding), and for all systems (with or without grounding.

This device includes:

- a protective device against indirect contact without grounding.

- a fault voltage detector without grounding.

1-Protective device against indirect contact (accidentally energized metallic body):

This device disconnects the machine and activates a led within a period of 10 to 20 ms after its detection of contact of $U_d=20 Vac$ and without grounding need.

The device voltage release $U_c=20 Vac$ is smaller than the standard upper safety limit $U_L=24 Vac$ for wet buildings.

This triggering voltage is not independent from default current which is the main origin of electric danger. The following formulas are considered.

May cause death

25mA in alternating current

The human body resistance is around 1000 Ω for dry skin.

There is danger for a voltage $U=0.025*1000=25$ Vac. Thus, the upper protective limit is set at $U_L=24 Vac$

For dry buildings and dry skin: $U_L=50 Vac$.

The device can be used in any system (with grounding, without grounding).

- For electrical systems without grounding : the protective device against indirect contact without grounding is intended to protect millions of users all over the world who don't have grounding in their electrical systems, and among them we find the highest number of electric choc casualties .
- For grounded electrical systems : the protective device against indirect contact without grounding is intended to minimize the hazards due to disadvantages of differential breakers and neutral regime , without any influence on the operation of all other protective devices being used.

-The outbreak of the device depends on the contact voltage Uc=20Vac which is harmless to human body by contact. On the other side, differential breaker having a sensitivity of 0.3A does not allow any protection if default current is less than 0.3A and the contact voltage is 230Vac (Such a default may cause death because a 20mA current causes diaphragm and respiratory muscles contraction, and a 30mA current provokes an important hazard of cardiac fibrillation).

- Problems that may occur on the grounding have no effect at all on the device operation.

The device allows locating the default and disconnects the failing machine as soon as the first default occurs.

-The device allows locating the default even in case a second default occurs on the same phase, which is almost impossible in neutral regime IT.

-Current disturbance does not provoke the device outbreak.

-The protective device against indirect contact (insulation default) without grounding is available in two models single phase and three phases.

-The device has no influence on the other devices and machines linked to the grounding system.

- For generators :

- Generators are an electric energy source for remote and off the grid sites. Accessibility to the generator's body and frequent contacts with it, as well as the impossibility to establish safe and permanent links to a reliable ground plug, require a big attention in order to make sure that in all cases users safety is assured. The device described hereafter solves this problem.

2-Default voltage detector without grounding:

- This detector monitors and locates the default without any breaking power.

- It can detect the default voltage as soon as it reaches 20Vac without the need of grounding and it activates a led or a warning system within a time between 10 and 20ms.

- The detector without grounding is intended to be used in industrial areas.

- It allows to quickly locating the default without the need of grounding which may diturb other protective devices.

- It has no effect on grounded devices and machines.

II-EXPLANATION

The invention operation principle is to detect the default voltage on the defective metallic machine body, and to use it to monitor an electromagnetic relay (automotive relay 12Vcc/35A).

- In case of insulation default, the contact voltage occurs as an electric current path sets up within the human body due to the voltage difference between the energized metallic machine body and the grounded human body.

- The invention presents a method that uses the default voltage without any grounding line to monitor electromagnetic relays through an NPN power transistor.

- The device includes three blocs:

Block1: Supply

This block is similar to a capacitive supply with few modifications. It is intended to rectify any voltage between 20 and 230Vac ==>0.006A < Is <0.07A

Diode zener 12V at supply block end

SCHEMA1

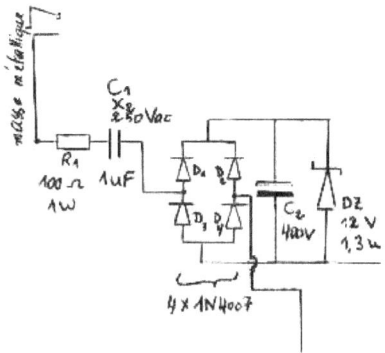

Pour Uc=220V   on a Uc= Z*I avec C= 1uF et C=50Hz   alors   I= U/Z= U*2π*f*c

AN: Is=0.07A=70mA

Pour Uc=20Vac   Is=0.006A=6mA

Block 2: Discharge fixed point

Whenever a voltage default sets up, this block discharges the capacitor X2 without the need to make the current flowing through the system neutral line.

SCHEMA2

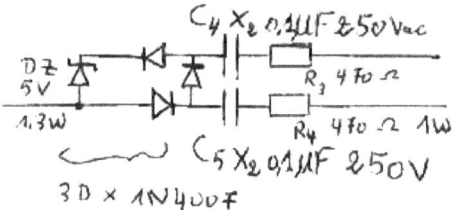

Is=0.005A=5mA needed by the LED..

- One can touch the discharge fixed point without any risk, because the current which is the hazard source is almost null.

Bloc 3: commande des relais electromagnetic

Schema 3

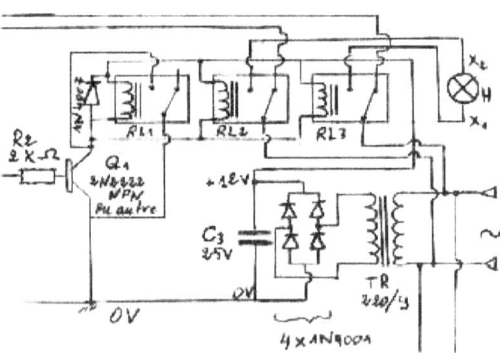

This block consists in :

- A power transistor NPN 2N2222

- A resistance R=2000Ω

Us=RI ➔ R=Us/I=12/0.006=2kΩ

Us: output voltage of Block 1

Is: output current of Block 1

- electromagnetic relays 1RT-NO/NF

Relays number is selected according to device type:

- Device for monophased machine : it has 3 relays , type 1RT-NO/NF
  - One retention relay to keep the coils excited after the tripping,
  -Two relays to insulate the machine and switch on warning light.
- Device for tri-phased machine : it has 4 relays, type 1RT-NO/NF
  - One retention relay to keep the coils excited after the tripping
  -Three relays to insulate the machine and to switch on the warning light or the alarm system.
- Default voltage detector : it has 2 relays, type 1RT-NO/NF
  - One retention relay to keep the coils excited after the tripping
  -One relay to switch on the warning light or the alarm system .

Protective device against electrical shock by indirect contact for single phased machines

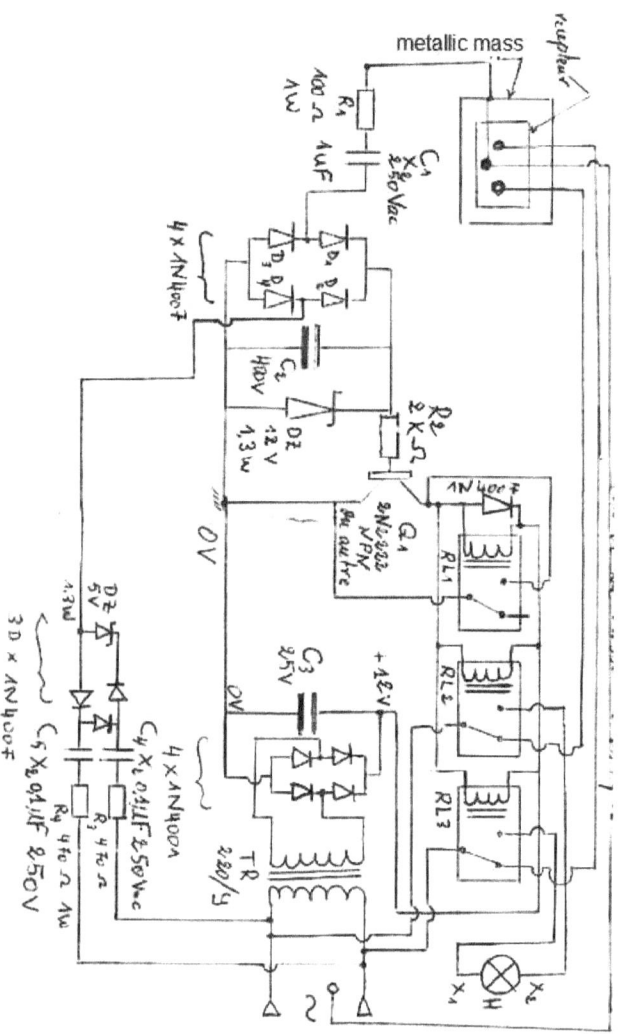

ABSTRACT

This invention is related to a protective device against indirect contact (accidentally energized metallic body) without grounding.

This device includes:

-a protective device against indirect contact (accidentally energized metallic body) without grounding

-a default voltage detector without grounding.

1- Protective device against indirect contact (accidentally energized metallic body) without grounding:

This device turns the machine off and switch on a warning light within a very short time between 10 and 20 ms as soon as a voltage contact is detected Ud=20Vac, and without the need the have a grounding line.

The device tripping voltage Uc=20Vac is smaller than the upper safety limit set by the standards for wet buildings UL=24Vac.

The device can be used in any electrical (with or without system grounding line).

- For systems without grounding line :

    he protective device against indirect contact without grounding is intended to protect millions of people all over the world using electrical systems without grounding line, among them high number of casualties are electrical shock hazards victims.

- For systems without grounding line

- the protective device without grounding is intended to minimize the electrical hazards due to malfunctions of differential circuit breakers and neutral regimes, while having no influence on the other protective devices operation.
- the tripping device depends on the contact voltage le Uc=20Vac which is not harmful for the human body, whereas a 0.3 A sensitivity differential circuit breaker has no protective role if the default current is smaller than 0.3A and the

contact voltage is 230Vac (Such a default may cause death because a 20mA current causes diaphragm and respiratory muscles contraction, and a 30mA current provokes an important hazard of cardiac fibrillation).
- Malfunctions occurring on the grounding line have no effect on the device operation and effectiveness.
- The device allows to locate the default and to turn the defective machine off as soon as the first default sets up.
- The device allows locating the default even in case of a second default on the same phase line, which is almost impossible with neutral regime IT.
- Minor voltage disturbance has no effect on the device and it doesn't make it tripping.
- The protective device against indirect contact (insulation default) without grounding is available in two models: single phased and tri-phased.

- For generators:

Generators are an electric energy source for remote and off the grid sites. Accessibility to the generator's body and frequent contacts with it, as well as the impossibility to establish safe and permanent links to a reliable ground plug, require a big attention in order to make sure that in all cases users' safety is assured. The device described in this document solves this problem

2-Voltage default detector without grounding

This detector has a monitoring and default locator function. It doesn't need any tripping power.

It is intended to be used in industrial systems without any effect on existing grounded machines and devices operation.

It allows:

-Default voltage detection as soon as it reaches 20Vac without the need of grounding, and it swithes on a warning light or an alarm system within a period of 10 to 20ms.

-Quick default location without the need of grounding line which can be a malfunction source for the other protective devices.

CLAIMS

1- Protective device against indirect contact (accidentally energized metallic body) without grounding, specified as follows:

- Protective device against electrical shocks caused by indirect contact with single phased machines without grounding.

- Protective device against electrical shocks caused by indirect contact with tri-phased machines without grounding.

- Default voltage detector for indirect contact without grounding.

2-Protective device against electrical shocks caused by indirect contact (accidentally energized metallic body due to insulation default) without grounding for single phased machines. It is intended to switch of the defective machine as soon as the voltage contact reaches 20Vac within a period of 10 to 20 ms. It can be used in any system (with or without grounding line) as per claim 1

3- Protective device against electrical shocks caused by indirect contact (accidentally energized metallic body due to insulation default) without grounding for tri- phased machines. It is intended to switch of the defective machine as soon as the voltage contact reaches 20Vac within a period of 10 to 20 ms. It can be used in any system (with or without grounding line) as per claim 1.

4-Voltage default detector during indirect contact (accidentally energized metallic body due to insulation default) without grounding. It is intended to locate the defective machine within a period of 10 to 20 ms, and it can be used for any kind of machines (single phased/tri-phased) and for any electrical system (with grounding line/without grounding line) as per claim 1.

www.ingramcontent.com/pod-product-compliance
Lightning Source LLC
Chambersburg PA
CBHW071146240526
45465CB00024BA/1796